A. De Morgan

AF874173

On the Calculation of Single Life Contingencies

Elibron Classics
www.elibron.com

Elibron Classics series.

© 2007 Adamant Media Corporation.

ISBN 1-4021-9146-4 (paperback)
ISBN 1-4021-2069-9 (hardcover)

This Elibron Classics Replica Edition is an unabridged facsimile of the edition published in 1840, London.

Elibron and Elibron Classics are trademarks of Adamant Media Corporation. All rights reserved.

This book is an accurate reproduction of the original. Any marks, names, colophons, imprints, logos or other symbols or identifiers that appear on or in this book, except for those of Adamant Media Corporation and BookSurge, LLC, are used only for historical reference and accuracy and are not meant to designate origin or imply any sponsorship by or license from any third party.

COMPANION TO THE ALMANAC

FOR

1840.

PART I.

INFORMATION CONNECTED WITH THE CALENDAR AND THE NATURAL PHENOMENA OF THE YEAR; AND WITH NATURAL HISTORY AND PUBLIC HEALTH.

I.—ON THE CALCULATION OF SINGLE LIFE CONTINGENCIES.

It is the object of the present article to put together a number of formulæ which it may be useful to the actuary to find in one place. At the same time, it may show all persons who possess an elementary knowledge of algebra, that they may, with no great amount of tables, and processes of very easy application, learn to compute the value of any benefit in which the duration of one life only is concerned. The same principles, with more extensive tables, apply to cases in which two or more lives are involved.

A sketch of the history of the subject will be found in the Library of Useful Knowledge, Treatise on *Probability;* and more fully in Mr. Milne's articles on *Mortality* and *Annuities* in the new edition of the Encyclopædia Britannica. The articles *Annuity, Interest, Mortality,* and *Reversions* (when the latter appears) in the Penny Cyclopædia, may also be consulted.

About thirty years ago, a Mr. George Barrett presented to the Royal Society a method by which the calculation of life contingencies was very materially facilitated. This method the society did not think worthy of publication; and it was accordingly given to the world by Mr. Francis Baily, in the appendix to his well-known work on Annuities, with some severe remarks on the omission just alluded to. It was certainly an unfortunate want either of examination or of judgment which caused the Philosophical Transactions, the depositary of the writings of many eminent inquirers on this particular subject, to miss a contribution which would have done honour to any one of them. This method of Mr. Barrett was rendered still more commodious, and we believe extended, by Mr. Griffith Davies, in his tables of life contingencies (1825), a work now unfortunately out of print: it is Mr. Barrett's method, as improved by Mr. Davies, which we propose to present to the reader, with some extension of notation and generalization of processes.

Let it be the law of mortality that of a_0 persons born alive, a_1 are alive at the end of a year, a_2 at the end of two years, a_x at the end of x years. Let v be the present value of £1, to be received at the end of a year, which depends entirely on the interest of money; if r be the interest of £1 for one year, we have

$$v = \frac{1}{1+r}:$$

thus, at 3 per cent., $r = \cdot 03$ and $v = \cdot 970874$. The present values of £1 to be received at the end of 2, 3, ... x years are v^2, v^3, ... v^x. The best way to find these powers will be to take the logarithms* from a larger table, such as that in the Penny Cyclopædia, article Interest. Thus the logarithm of $1+r$ being

$$\cdot 0128372247, \text{ that of } v \text{ is } 9\cdot 9871627753 - 10$$

Multiples of this logarithm being formed up to 104 times, we have all the logarithms which will be wanted, in the use of the Carlisle Table.

Persons not used to computation should remember that the easiest way of forming a set of multiples is to write the quantity to be added each time at the bottom of a card, and to make each addition by holding the card so that the writing on it may stand over the last result. In this way it will not take many minutes to form a hundred multiples of the preceding, and a verification of the last multiple should be made by actual multiplication.

Obtain as many logarithms as are wanted of the powers of v in the preceding manner, allowing as much space between the lines as will† contain four lines of figures. Take the table of mortality which is to be used, say the Carlisle table, and under the logarithm of v write that of a_1, the number surviving a year; under that of v^2, write the logarithm of a_2; and so on up to the end of life. The Carlisle table will be found in Mr. Milne's work on Annuities; it is also in the Penny Cyclopædia, article‡ *Mortality.* Under the last logarithms, write in succession the logarithms of $a_0 - a_1$, $a_1 - a_2$, &c., the numbers who die in the first, second, &c. years: as follows, in which 3 per cent. is supposed. Five decimal places are taken, merely as an example; but it will be almost as easy to use seven, as there are no interpolations.

	10000				
1		$\log v$	$= 9\cdot 98716 - 10$		
2	8461	$\log a_1$	$= 3\cdot 92742$		
3	1539	$\log(a_0 - a_1)$	$= 3\cdot 18724$		
4			$3\cdot 91458$	$a_1 v$	$= 8214\cdot 5$
5			$3\cdot 17440$	$(a_0 - a_1)v$	$= 1494\cdot 2$
1		$\log v^2$	$= 9\cdot 97433 - 10$		
2	7779	$\log a_2$	$= 3\cdot 89092$		
3	682	$\log(a_1 - a_2)$	$= 2\cdot 83378$		
4			$3\cdot 86525$	$a_2 v^2$	$= 7332\cdot 5$
5			$2\cdot 80811$	$(a_1 - a_2)v^2$	$= 642.85$
	&c.	&c.	&c.	&c.	&c.

* We have re-examined this table, and find no error, by Hutton's Tables, p. 386.]
† It will be desirable to have the spaces equal.
‡ We have re-examined this reprint, and find it correct.

Those who cannot easily add one line to another which is separated by a third should now cut pieces out of a card in such manner that being laid upon one of the compartments of the table, the parts cut open will show the first and third line, and the rest hide the second and fourth. The fourth line is then formed by adding the first and second, and the fifth line by adding the first and third, covering the second and fourth. We thus obtain two successions of results.

$$a_0 \qquad a_1 v \qquad a_2 v^2 \qquad a_3 v^3, \qquad \&c.$$

$$(a_0 - a_1)v \qquad (a_1 - a_2)v^2 \qquad (a_2 - a_3)v^3 \qquad (a_3 - a_4)v^4, \qquad \&c.$$

Let these be denominated D_0, D_1, D_2, &c., and C_0, C_1, C_2, &c., so that

$$D_0 = a_0, \qquad D_1 = a_1 v, \ldots\ldots\ldots D_x = a_x v^x$$

$$C_0 = (a_0 - a_1)v, \quad C_1 = (a_1 - a_2)v^2, \ldots\ldots C_x = (a_x - a_{x+1})v^{x+1}.$$

The following table is then to be completed in the manner which will be described.

Age.	D_x	N_x	S_x	C_x	M_x	R_x	Age
0	D_0	N_0	S_0	C_0	M_0	R_0	0
1	D_1	N_1	S_1	C_1	M_1	R_1	1
2	D_2	N_2	S_2	C_2	M_2	R_2	2
3	D_3	N_3	S_3	C_3	M_3	R_3	3
&c.	&c.	&c.	&c.	&c.	&c.	&c.	&c.

The columns D and C have been described; the rest are formed from them as follows. The last of the column N is nothing, and N_x is always the sum of those in column D beginning with D_{x+1}, and continuing to the end: thus any one D and its N added together give the preceding N. Or

$$N_x = D_{x+1} + D_{x+2} + D_{x+3} + \ldots\text{(to the end)}$$

$$N_x = D_{x+1} + N_{x+1}.$$

The last S is nothing, and the column S is formed from N as N was formed from D in every point except this, that each S begins with its own N instead of the one after: thus

$$S_x = N_x + N_{x+1} + N_{x+2} + \ldots\text{(to the end)}$$

$$S_x = N_x + S_{x+1}.$$

The last M is the last C, and M is formed from C precisely as S from N: thus

$$M_x = C_x + C_{x+1} + C_{x+2} + \ldots\text{(to the end)}$$

$$M_x = C_x + M_{x+1}.$$

Lastly, R is formed from M as M from C: thus

$$R_x = M_x + M_{x+1} + M_{x+2} + \ldots\text{(to the end)}$$

$$R_x = M_x + R_{x+1}.$$

We here give, as a specimen, the first five, the last five, and an intermediate five years, from the Carlisle table at 3 per cent., keeping only four significant figures.

Age.	D.	N.	S.	C.	M.	R.	Age	No. Living	Dying.
0	10000	173200	3702000	1494·0	4664	70040	0	10000	1539
1	8215	165000	3529000	642·9	3170	65372	1	8461	682
2	7332	157700	3364000	462·1	2527	62202	2	7779	505
3	6657	151000	3206000	245·2	2065	59674	3	7274	276
4	6218	144800	3055000	173·4	1820	57610	4	6998	
·	·	·	·	·	·	·	·	·	·
30	2324	45460	732100	22·80	932·7	25070	30	5642	57
31	2234	43220	686600	22·14	909·9	24140	31	5585	57
32	2147	41080	643400	21·11	887·8	23230	32	5528	56
33	2063	39020	602300	20·13	866·6	22340	33	5472	55
34	1983	37030	563300	19·55	846·5	21470	34	5417	
·	·	·	·	·	·	·	·	·	·
100	·4683	·7879	1·4580	·10100	·43170	1·17700	100	9	2
101	·3536	·4343	·6696	·09809	·33070	·74550	101	7	2
102	·2452	·1891	·2353	·09524	·23260	·41480	102	5	2
103	·1429	·0462	·0462	·09246	·13730	·18220	103	3	2
104	·0462	·0000	·0000	·04488	·04488	·04488	104	1	

That the beginner may attach definite ideas to the several columns, we subjoin an explanation of each. The assumption is that of 10000 individuals born alive, the numbers surviving to each age, and dying in each year, are as in the last two columns.

D. £2324 invested at 3 per cent. at the birth of 10,000 persons, will, improved at compound interest, yield every survivor £1 at the age of 30: the number of survivors being 5642.

N. £45460 invested at the birth of 10,000 individuals will produce, by the time they attain the age of 30, enough to guarantee to each person then surviving an annuity of £1 on his life, the first payment being made when he attains the age of 31.

S. £732,100 will in the same case produce enough to guarantee an increasing annuity, paying £1 to each at the age of 31, £2 at that of 32, &c.

C. £22·8 invested at 3 per cent. at the birth of 10,000 persons, will, improved at compound interest until the survivors are 31 years of age, yield £1 for each of those who died between 30 and 31.

M. £932·7 similarly invested, would yield to each one who reaches 30, £1 at the end of the year in which he dies.

R. £25070 similarly invested, would yield to each one who

lives to be 30, £1 if he die in the 31st year, £2 if in the 32nd, and so on.

The preceding list simply enunciates the method of constructing the tables; the following shows the use to which the mere inspection may be put. Take any two ages, say 30 and 34, and transpose the numbers opposite each age to the other age: then, whatever may be the present age (less than 30);—

D. A person might now give up £1983, due at the age of 30, to receive £2324, if he live to be 34.

N. A person might now give up an annuity of £37030, to be granted at the age of 30, to receive in return another of £45460, to be granted at the age of 34, if he should live so long.

S. A person might now give up a uniformly increasing annuity of £563,300 the first year, twice as much the second, &c. to be entered upon at the age of 30, to receive another annuity of the same kind, beginning with £732,100, to be entered upon at the age of 34, if he should live so long.

C. £19·55 secured to a person in the event of his dying between 30 and 31, is now of the same value as £22·80, secured to the same person in the event of his dying between 34 and 35.

M. £846·5 secured to a person at the end of the year in which he dies, if after attaining 30, is of the same value as £932·7 secured to the same person at the same period, if after attaining 34.

R. An increasing assurance, to be £21470, if a person die in his 31st year, twice as much if in his 32nd, &c., is now of the same value as another, to be £25070 if he die in his 35th year, twice as much if in his 36th, &c.

These properties are independent of the present age of the party, and show that the most simple indication of the tables is the proportion in which a benefit due at one age ought to be changed, so as to retain the same value and be due at another age. They might, therefore, with great propriety be called *commutation tables*.

The following formulæ and additional notation will be found useful.

I. $N_x = D_{x+1} + D_{x+2} + D_{x+3} + \dots$ to the end of life.

$M_x = C_x + C_{x+1} + C_{x+2} + \dots \quad . \quad . \quad . \quad . \quad .$

$S_x = N_x + N_{x+1} + N_{x+2} + \dots = D_{x+1} + 2D_{x+2} + 3D_{x+3} + \dots$

$R_x = M_x + M_{x+1} + M_{x+2} + \dots = C_x + 2C_{x+1} + 3C_{x+2} + \dots$

II. $N_x - N_{x+1} = D_{x+1}$ $\qquad$ $S_x - S_{x+1} = N_x$

$M_x - M_{x+1} = C_x$ $\qquad$ $R_x - R_{x+1} = M_x$

III. Let $N_{x,y} = N_x - N_{x+y}$ $\quad|\quad$ $S_{x,y} = S_x - S_{x+y} - yN_{x+y-1}$

$M_{x,y} = M_x - M_{x+y}$ $\quad|\quad$ $R_{x,y} = R_x - R_{x+y} - yM_{x+y-1}$

IV. Then $N_{x,y} = D_{x+1} + D_{x+2} + \ldots + D_{x+y}$ ...(y terms)

$M_{x,y} = C_x + C_{x+1} + \ldots + C_{x+y-1}$ ...(y terms)

$S_{x,y} = D_{x+1} + 2D_{x+2} + \ldots + (y-1)\,D_{x+y-1}$ ...($y-1$ terms)

$R_{x,y} = C_x + 2C_{x+1} + \ldots + (y-1)\,C_{x+y-2}$ ...($y-1$ terms).

V.

$$D_x + \ldots + D_y = N_{x-1,\,y-x+1}$$
$$C_x + \ldots + C_y = M_{x,\,y-x+1}$$
$$D_x + \ldots + (y-x+1)\,D_y = S_{x-1,\,y-x+2}$$
$$C_x + \ldots + (y-x+1)\,C_y = R_{x,\,y-x+2}$$

VI.

$$N_{x,y} + S_{x+1,y} = S_{x,y+1} = S_x - S_{x+y} - yN_{x+y}$$
$$M_{x,y} + R_{x+1,y} = R_{x,y+1} = R_x - R_{x+y} - yM_{x+y}$$

VII.

$$N_{x,y} - \frac{1}{y}S_{x+1,y} = N_x - \frac{1}{y}(S_{x+1} - S_{x+y+1})$$
$$M_{x,y} - \frac{1}{y}R_{x+1,y} = M_x - \frac{1}{y}(R_{x+1} - R_{x+y+1})$$

VIII.

$$C_x = v\,D_x - D_{x+1}, \quad M_x = v\,N_{x-1} - N_x$$
$$R_x = v\,S_{x-1} - S_x$$
$$M_{x,y} = v\,N_{x-1,y} - N_{x,y}, \quad R_{x,y} = v\,S_{x-1,y} - S_{x,y}$$

The formulæ VIII. will be useful in verifying the tables.

All that will be found of demonstration in the present article is intended for those who are familiar with the subject, being meant to give a method of dealing with the more complicated cases, and particularly a method by which the succeeding formulæ may be verified. This will be followed by a collection of preparations for formulæ which may be easily used even by a person unacquainted with the demonstration.

The present value of £1 to be received by a person now aged x, if he live to attain $x+k$, is $D_{x+k} \div D_x$.

The present value of £1 to be received by the representatives of a person now aged x, if he die between the ages of $x+k$ and $x+k+1$, is $C_{x+k} \div D_x$.

The following problem will include every case we have yet seen proposed of annuities, whether for the whole life, or temporary, or deferred, increasing or decreasing uniformly; and also of insurances: with every manner yet proposed of paying the premium. It matters nothing that it involves payment of premium after the benefits begin to be received; since every application of it will require the part of the premium so paid to be made equal to nothing.

PROBLEM. A person now aged x is to receive s (pounds sterling) if he attain $x+k$. In the n following years he is, if he live, to receive a, $a+h$, ... $a+(n-1)\,h$ pounds at the end of the successive years, and ever afterwards during his life, t pounds at the end of each year. But if he die during the n years, he is to have A, A+H .... A+$(n-1)$ H, according as he dies in the first, second, &c. year, and T at the end of any subsequent year in which he dies. Besides this, he is, if he die in w years, to have a return of part of the premiums presently described.

For this he is to pay at once, σ, and a premium ϖ, which he is to pay l times; the next m premiums are to be ϖ, $\varpi(1+\mu)+\beta$, $\varpi(1+2\mu)+2\beta, \ldots . \varpi(1+\overline{m-1}\,\mu)+(m-1)\,\beta$; after which the premium is always τ. But if he should die before he attains $x+w$ years, he is to receive at the end of the year of his death $\varrho+\nu\varpi$ if he die in the first year, $\varrho+\theta+2\nu\varpi$ if in the second, and, finally, $\varrho+(w-1)\,\theta+w\nu\varpi$ if in the wth year. Required the equation that must exist among these quantities to make the receipts and payments of equal value.

The first set of receipts has the value of the following expression divided by D_x.

$$\begin{aligned}&aD_{x+k+1}+(a+h)\,D_{x+k+2}+\ldots\ldots+(a+\overline{n-1}\,h)\,D_{x+k+n}\\&+AC_{x+k}+(A+H)\,C_{x+k+1}+\ldots .+(A+\overline{n-1}\,h)\,C_{x+k+n-1}\\&+sD_{x+k}+t\,(D_{x+k+n+1}+\ldots .)\quad +T\,(C_{x+k+n}+\ldots .),\end{aligned}$$

which, by the preceding formulæ, is

$$\left\{\begin{matrix}aN_{x+k,n}+hS_{x+k+1,n}+AM_{x+k,n}+HR_{x+k+1,n}\\ +sD_{x+k}+tN_{x+k+n}+TM_{x+k+n}\end{matrix}\right\}\ldots .(A).$$

The balance of the premiums and returns is the following divided by D_x:

$$\begin{aligned}&(\sigma+\varpi)\,D_x+\varpi\,(D_{x+1}+\ldots+D_{x+l-1})+\varpi D_{x+l}+(\varpi\,\overline{1+\mu}+\beta)\,D_{x+l+1}\\&+\ldots .+\{\varpi\,(1+\overline{m-1}\,\mu)+\overline{m-1}\,\beta\}\,D_{x+l+m-1}+\tau\,(D_{x+l+m}+\ldots .)\\&-(\varrho+\nu\varpi)\,C_x-(\varrho+\theta+2\nu\varpi)\,C_{x+1}-\ldots-(\varrho+\overline{w-1}\,\theta+w\nu\varpi)\,C_{x+w-1},\end{aligned}$$

the value of which, by the same formulæ, is

$$\left\{\begin{matrix}\sigma D_x+\varpi N_{x-1,l+m}+\varpi\mu S_{x+l,m}+\beta S_{x+l,m}+\tau N_{x+l+m-1}\\ -\varrho M_{x,w}-\theta R_{x+1,w}-\nu\varpi R_{x,w+1}\end{matrix}\right\}\ldots .(B).$$

The equation of the results (A) and (B) gives

$$\left.\begin{matrix}\varpi\{N_{x-1,l+m}+\mu S_{x+l,m}-\nu R_{x,w+1}\}\\ +\sigma D_x+\beta S_{x+l,m}+\tau N_{x+l+m-1}\\ -\rho M_{x,w}-\theta R_{x+1,w}\end{matrix}\right\}=\left\{\begin{matrix}sD_{x+k}\\ +aN_{x+k,n}+hS_{x+k+1,n}\\ +AM_{x+k,n}+HR_{x+k+1,n}\\ +tN_{x+k+n}+TM_{x+k+n}\end{matrix}\right.$$

This problem contains the circumstances of all which are proposed, and is here introduced that any question may have the result of the common investigation compared with that deduced by considering it as a particular case of the preceding. Most of the letters will be $=0$ in any question which occurs; and the following list will serve to remind the calculator what letters enter into the case before him.

1	k	s	Occurs in questions which comprise a fixed sum at a certain age; an endowment.
2	k, n	a	An annuity, present or deferred, &c.
3	k, n	h	An increasing or decreasing annuity.
4	k, n	A	A fixed assurance of any kind.
5	k, n	H	An increasing or decreasing assurance.
6	k, n	t	An increasing or decreasing annuity, with an end to the increase, &c.
7	k, n	T	An increasing or decreasing assurance with an end to the increase, &c.
8	—	σ	Present value of any kind.
9	l, m, w	ϖ	Fixed premiums of any kind.
10	l, m	μ	Premiums increasing or diminishing by a proportion of the first premium.
11	l, m	β	Premiums increasing or decreasing by a sum independent of the first premium.
12	l, m	τ	Premiums increasing or decreasing with an end to the increase, &c.
13	w	ϱ	Return of a sum in case of death.
14	w	θ	Return of increasing or decreasing absolute sum in case of death.
15	w	ν	Return of increasing or decreasing proportions of the first premium in case of death.

The third column contains the letters indicating benefits or payments, and the second column shows the terms of years with which the benefits, &c. are particularly connected in the general problem. By attention to the conditions, the solution of any case can be readily picked out of the general equation, as in the following instances.

What is the premium to be paid for an insurance of £A on a life aged x, accompanied by a return at death of all the premiums paid?

Here all the letters of the third column vanish except A, ϖ, and ν, and m, n, and w are to be extended beyond the possible term of life, while $l=0$, $k=0$, and $\nu=1$. Again, when the age $x+y$ extends beyond the term of life, $N_{x,y} = N_x$, &c. Consequently, the equation gives

$$\varpi(N_{x-1} - R_x) = AM_x.$$

Again, what is the present value of an assurance of £A, with which the sum paid is to be returned? Here only σ, ϱ, and A have value, and $\sigma=\varrho$, the letters of the first column being as before. Hence

$$\sigma(D_x - M_x) = AM_x.$$

If D_x be ever less than M_x this problem is impossible· but D_x is necessarily greater than M_x, being $C_x(1+r)+C_{x+1}(1+r)^2+\ldots$, while M_x is $C_x+C_{x+1}+\ldots$.

Thirdly, an annuity of £a to commence when a life aged x attains $x+k$ is to be bought by a premium regularly diminishing, so as to be last paid when the annuity begins (that is at $x+k$), and a year before a payment of the annuity is made. Here only ϖ, μ, and a have value, k is given, $l=0$, m is $k+1$, n outruns the term of life, and $w=0$. And μ must be taken negatively, as $-1\div(k+1)$. Hence

$$\varpi\left(N_{x-1\,k+1,}-\frac{1}{k+1}S_{x,\,k+1}\right)=a.N_{x+k},$$

or (VII.) $$\varpi=\frac{aN_{x+k}}{N_{x-1}-(S_x-S_{x+k+1})\div(k+1)}.$$

Before we proceed further, it may increase the interest attached to the formulæ if we remark that the principle of these commutation tables (as we have called them) can be extended from the case of life contingencies to that of interests certain, in such manner that every formula which gives the value of one of the former, may, by going to a different table, be applied to the corresponding one of the latter. That is to say, the mathematical treatment of the hypothesis that a life is to last for ever does not differ from that of a table of *mortality*. We should imagine, that in questions of instalments particularly, increasing or decreasing, such tables would be of very great use.

To construct them, proceed as follows:

$$D_x=v^x,\quad N_x=\frac{v^{x+1}}{1-v},\quad S_x=\frac{v^{x+1}}{(1-v)^2}.$$

The remaining quantities are useless, being always $=0$. $N_{x,y}$ and $S_{x,y}$ may be exhibited as before in the forms

$$N_{x,y}=N_x-N_{x+y},\quad S_{x,y}=S_x-S_{x+y}-yN_{x+y-1}.$$

Suppose, for instance, we take the last question, and require the value of the annuity for the *whole life* (which here means perpetuity) of a after the expiration of k years, to be bought by regularly diminishing instalments one paid now, &c. The formula then becomes

$$\varpi=\frac{aN_k}{N_{-1}-(S_0-S_{k+1})\div\overline{k+1}},$$

and $N_{-1}=\dfrac{1}{1-v}$, $S_0=\dfrac{v}{(1-v)^2}$. If the value of the tabular quantities be restored, the preceding (cleared of fractions) is

$$\frac{a\,(k+1)\,v^{k+1}\,(1-v)}{(k+1)\,(1-v)-(v-v^{k+2})};$$

the same as would be obtained by common methods. The first is (with tables) as easily calculated as the second by the common tables; or, if any thing, somewhat more easily.

The particular uses of such commutation tables for certain interests would be, 1. That those who can use the life tables more readily than the usual tables of interest (many, perhaps most, actuaries) would at once be able to apply their facility in the former to the new form of the latter. 2. That whenever the formula is given in terms of D_y, N_y, and S_y, it is indifferent at what age the perpetual life of the problem is supposed to begin, so that a repetition of the simple process upon another age verifies the computation.

We now come to the classification of problems, and the presentation of their results. In all cases, one of the quantities in the third column is to be unknown, and found from the equation. Two cases arise. 1. Where all the quantities of the third column are independent of each other. 2. Where one is to be a simple fraction of another. Thus τ, the premium remaining over, or what we might call the *residual* premium, might be required to be $\gamma\varpi$, a given fraction of ϖ; and the problem might as easily be solved if all except, say a, (and the indicators of fractions already existing, as μ, ν) were to be made given fractions or multiples of ϖ.

If $(\nu\varpi)$, in parentheses, be taken as an abbreviation of the phrase 'coefficient of $\nu\varpi$,' &c., we may, for purposes of general consideration, write the equation

$$\varpi\left\{(\varpi)+(\mu\varpi).\mu-(\nu\varpi).\nu\right\}+\&c. = (s).s+(a).a+\&c.$$

the first side entirely depending upon the mode of offering payment, and the second upon the nature and amount of the benefit. Hence it is useless to combine each of the different benefits with all the modes of paying for it; for, as cannot but have been observed by those who have used these tables, a given benefit must be always calculated by the same numerator, whatever the single mode of payment may be; and the payment by the same denominator, whatever the benefit may be. Thus if a simple deferred annuity be bought by uniform premiums, we have

$$(\varpi).\varpi=(a).a, \quad \text{or} \quad \varpi=\frac{(a)}{(\varpi)}.a.$$

But if the single value be paid for the same, we have

$$(\sigma)\,\sigma=(a).a, \quad \text{or} \quad \sigma=\frac{(a)}{(\sigma)}.a.$$

But if the mode of payment be double, say partly by a single value, partly by a succession of uniform premiums, we have

$$(\varpi).\varpi+(\sigma)\,\sigma=(a).a;$$

from which ϖ may be given, and σ be found, or vice versâ. If σ is to be a given fraction of ϖ, the *payment* part of the calculation is wholly in the denominator. The following rules will be found useful as preservatives from error.

1. When no part of the benefit is to depend upon the unknown

item of payment,* no function of the benefit can be in the denominator ; and the contrary.

2. When all the items of payment are fractions of one among them, no function of the payment can be in the numerator, and the contrary : but when there are parts of the payment not so connected, those which are known are found in the numerator.

3. In all the cases not before specified, the numerator is entirely a function of the benefit and the denominator of the payment.

If we call the two sides of the equation the payment side and the benefit side, and if taking twenty different cases of each, we write down the corresponding sides of the equation, we have the materials for solving instantaneously any one out of four hundred problems, out of which all will be practically useful, in which the conditions of the problem make all payments cease at or before the time when the benefit begins. This we proceed to do. The equation is

$$\left.\begin{array}{c}\sigma D_x - \varrho M_{x,w} \\ +\varpi\{N_{x-1,l+m} + \mu S_{x+l,m} - \nu R_{x,w+1}\} \\ +\tau N_{x+l+m-1} + \beta S_{x+l,m} - \theta R_{x+1,w}\end{array}\right\} = \left\{\begin{array}{c} sD_{x+k} \\ + aN_{x+k,n} + hS_{x+k+1,n} + tN_{x+k+n} \\ + AM_{x+k,n} + HR_{x+k+1,n} + TM_{x+k+n}\end{array}\right.$$

$$N_{x,y} = N_x - N_{x+y} \qquad S_{x,y} = S_x - S_{x+y} - yN_{x+y-1}$$
$$M_{x,y} = M_x - M_{x+y} \qquad R_{x,y} = R_x - R_{x+y} - yM_{x+y-1}$$

The following are the principal cases of benefits to be bought, and under each is written the benefit side of the equation in which it enters. The age of the life is x throughout.

Benefit Terms. Annuities.

1. *Endowment.* £s to be received in k years if the party be then alive,

$$sD_{x+k}\,.$$

2. *Life Annuity* of £a. First payable in one year, continuing through life,

$$aN_x\,.$$

3. *Deferred Life Annuity.* Deferred for k years, makes payment in $k+1$ years,

$$aN_{x+k}\,.$$

4. *Temporary Annuity.* Makes no payment after n years though the annuitant continue alive,

$$a\,(N_x - N_{x+n}).$$

5. *Temporary deferred Annuity.* Deferred k, continues n, years, first payment in $k+1$ years,

$$a\,(N_{x+k} - N_{x+k+n}).$$

* With the payment class any returns of payment in case of the conditions of benefit ceasing to exist before it becomes due.

6_1. *Increasing or decreasing Life Annuity.* Differs from (2) in the successive payments being a, $a \pm h$, $a \pm 2h$, &c.,

$$a\mathrm{N}_x \pm h\mathrm{S}_{x+1}.$$

6_2. When $h = a$ for the *increasing* annuity,*

$$a\mathrm{S}_x.$$

7_1. *Deferred increasing or decreasing Annuity.* Deferred k years, first payment a, in $k+1$ years, second $a \pm h$ in $k+2$ years, &c.

$$a\mathrm{N}_{x+k} \pm h\mathrm{S}_{x+k+1}.$$

7_2. When $h = a$, for the *increasing* annuity,

$$a\mathrm{S}_{x+k}.$$

8_1. *Temporary increasing or decreasing Annuity.* Lasts n years only.

$$a(\mathrm{N}_x - \mathrm{N}_{x+n}) \pm h(\mathrm{S}_{x+1} - \mathrm{S}_{x+n+1} - n\mathrm{N}_{x+n})$$

8_2. When $h = a$, for the *increasing* annuity,

$$a(\mathrm{S}_x - \mathrm{S}_{x+n} - n\mathrm{N}_{x+n}).$$

9_1. *Temporary deferred increasing or decreasing Annuity.* Deferred k years, continues n years,

$$a(\mathrm{N}_{x+k} - \mathrm{N}_{x+k+n}) \pm h(\mathrm{S}_{x+k+1} - \mathrm{S}_{x+k+n+1} - n\mathrm{N}_{x+k+n}).$$

9_2. When $h = a$, for the *increasing* annuity,

$$a(\mathrm{S}_{x+k} - \mathrm{S}_{x+k+n} - n\mathrm{N}_{x+k+n}).$$

10. *Decreasing Annuity, temporary by extinction.* That is, it lasts n years, and each payment is less than the preceding by 1-nth part of a $(hn = a)$,

$$a\left\{\mathrm{N}_x - \frac{1}{n}(\mathrm{S}_{x+1} - \mathrm{S}_{x+n+1})\right\}.$$

11. *Deferred decreasing Annuity, temporary by extinction.* Deferred k years, expires after $k+n$ years,

$$a\left\{\mathrm{N}_{x+k} - \frac{1}{n}(\mathrm{S}_{x+k+1} - \mathrm{S}_{x+k+n+1})\right\}.$$

12. *Arrested increasing or decreasing Annuity.* Here after n years, when the annuity would begin to pay $a \pm nh$, $a \pm (n+1)h$, &c., the increase or decrease is *arrested*, and it pays $a + (n-1)h$ for the rest of life,

$$a\mathrm{N}_x \pm h(\mathrm{S}_{x+1} - \mathrm{S}_{x+n}).$$

13. *Increasing or decreasing Annuity, deferred and arrested.* The period of deferment is k years, and the increase or decrease continues n years, after which as in the last,

$$a\mathrm{N}_{x+k} \pm h(\mathrm{S}_{x+k+1} - \mathrm{S}_{x+k+n}).$$

14_1. *Temporary Annuity, continued by increase or decrease.* Here the annuity is a for $n+1$ years,† after which it increases or decreases by h for $p-1$ years, and then stops.

$$a(\mathrm{N}_x - \mathrm{N}_{x+n+p}) \pm h(\mathrm{S}_{x+n+1} - \mathrm{S}_{x+n+p+1} - p\mathrm{N}_{x+n+p}).$$

* This case can be easily calculated from the common life tables, by a method given by the author of this article in his Essay on Probabilities. (Cabinet Cyclopædia.)

† Namely, for n years from the first part, and one year of the continuation.

14_2. The same *continued to extinction*, $(hp = a)$.

$$a\left\{N_x - \frac{1}{p}(S_{x+n+1} - S_{x+n+p+1})\right\}.$$

15_1. *Deferred temporary Annuity, continued by increase or decrease.* Here, after k years, a is paid for $n+1$ years, and $a \pm h$, $a \pm 2h$, &c. during $p-1$ years more,

$$a(N_{x+k} - N_{x+k+n+p}) \pm h(S_{x+k+n+1} - S_{x+k+n+p+1} - pN_{x+k+n+p}).$$

15_2. The same *continued to extinction*, $(hp = a)$.

$$a\left\{N_{x+k} - \frac{1}{p}(S_{x+k+n+1} - S_{x+k+n+p+1})\right\}.$$

In the preceding list it is obvious that any benefit there described is converted into another of the same kind, but deferred for k years, simply by changing x into $x+k$. In like manner, the benefit might be anticipated a year, by writing $x-1$ for x, which would make all the immediate annuities become due, or would alter their technical character from *annuities* to *premiums*. Similarly, if we compare the meanings of $D_{x+k} \div D_x$ and $C_{x+k} \div D_x$, we see that

The first is the value of £1, to be received if the person *begin* his $(x+k+1)$th year, whether he live through it or not.	The second is the value of £1, to be received if the person *begin* his $(x+k+1)$th year, *and do not live to finish it.*

If, then, we change D_x into C_x, &c., in any problem of annuities, and alter the benefit side of the equation accordingly, we make a change of benefits as follows: at every period at which the claimant, being alive, should receive a sum of £1, let him receive it a year later, but only if he die within the year, and let it be forfeited if he live. Consequently, an annuity to be paid, say at the seventh, eighth, and ninth birthday from the present time, would thus be turned into an assurance to be paid at the eighth, ninth, or tenth birthday, if the party should die in either of these years. But since M_x was made to begin a year earlier than N_x, or $M_x = C_x + \ldots$ and $N_x = D_{x+1} + \ldots$, this change of conditions as to time is compensated by the structure of the tables; and any one of the preceding annuity benefits is converted into its corresponding assurance benefit, so far as the benefit side of the equation is concerned, by changing N_x into M_x and S_x into R_x. But if D_x ever occur, we must change it into C_{x-1} if the time of payment is to be the same in both.

We might thus dispense with the following list, but, in an article of reference, it is desirable, were it only to avoid the necessity of looking under one head, while thinking of another.

Benefit Terms. Assurances.

1. *Endowment Assurance.* £S to be received in k years, if the person now aged x died in the preceding year,

$$SC_{x-1}.$$

2. *Life Assurance* of £A.* Payable at the end of the year of death,

$$AM_x.$$

3. *Deferred Assurance.* Payable at death, if more than k years hence,

$$AM_{x+k}.$$

4. *Temporary Assurance.* Payable at death if within n years,

$$A(M_x - M_{x+n}).$$

5. *Temporary deferred Assurance.* Payable at death, if between the ages of $x+k$ and $x+k+n$,

$$A(M_{x+k} - M_{x+k+n}).$$

6_1. *Increasing or decreasing Life Assurance.* Payable at death, A if in the first year, A $\pm$ H if in the second, A $\pm$ 2H if in the third, &c.

$$AM_x \pm HR_{x+1}.$$

6_2. When H = A, for the *increasing* assurance.

$$AR_x.$$

7_1. *Deferred increasing or decreasing Assurance.* Deferred k years, A if death in $(k+1)$th year, &c.

$$AM_{x+k} \pm HR_{x+k+1}.$$

7_2. When H = A for the *increasing* assurance

$$AR_{x+k}.$$

8_1. *Temporary increasing or decreasing Assurance.* If death take place in n years,

$$A(M_x - M_{x+n}) \pm H(R_{x+1} - R_{x+n+1} - nM_{x+n}).$$

8_2. When H = A for the *increasing* assurance,

$$A(R_x - R_{x+n} - nM_{x+n}).$$

9_1. *Temporary deferred increasing or decreasing Assurance.* Deferred k years, continues n years,

$$A(M_{x+k} - M_{x+k+n}) \pm H(R_{x+k+1} - R_{x+k+n+1} - nM_{x+k+n}).$$

9_2. When H = A for the *increasing* assurance,

$$A(R_{x+k} - R_{x+k+n} - nM_{x+k+n})$$

10. *Decreasing Assurance, temporary by extinction.* Payable at death, A, if in the first year, $(n-1)$-$n$ths of A if in the second, &c.,

$$A\left\{M_x - \frac{1}{n}(R_{x+1} - R_{x+n+1})\right\}.$$

11. *Deferred decreasing Assurance, temporary by extinction.* Payable A, if death take place in the $(k+1)$th year, &c.,

$$A\left\{M_{x+k} - \frac{1}{n}(R_{x+k+1} - R_{x+k+n+}).\right.$$

* Actuaries say *assurance*, and others *insurance*. The difference may be made useful in remembering (what the courts of law have not yet found out) that a life *assurance* and a fire *insurance* are very different things.

12. *Arrested increasing or decreasing Assurance.* Here, after n years, when the sum payable should be $A \pm nH$, &c., the increase or decrease is arrested, and $A \pm (n-1)H$ is the assurance for the rest of life.

$$AM_x \pm H(R_{x+1} - R_{x+n}).$$

13. *Increasing or decreasing Assurance, deferred and arrested.* Deferment k years, increase or decrease n years, after which as in the last.

$$AM_{x+k} \pm H(R_{x+k+1} - R_{x+k+n}).$$

14_1. *Temporary Assurance, continued by increase or decrease.* Here the assurance is A for $n+1$ years, after which it increases or decreases by H for $p-1$ years, and then stops,

$$A(M_x - M_{x+n+p}) \pm H(R_{x+n+1} - R_{x+n+p+1} - pM_{x+n+p}).$$

14_2. The same *continued to extinction*, $(Hp = A)$,

$$A\left\{M_x - \frac{1}{p}(R_{x+n+1} - R_{x+n+p+1})\right\}.$$

15_1. *Deferred temporary Assurance, continued by increase or decrease.* Here, after k years, A is the assurance for $n+1$ years, and $A \pm H$, $A \pm 2H$, &c. for $p-1$ years more.

$$A(M_{x+k} - M_{x+k+n+p}) \pm H(R_{x+k+n+1} - R_{x+k+n+p+1} - pM_{x+k+n+p}).$$

15_2. The same, *continued to extinction*, $(Hp = A)$,

$$A\left\{M_{x+k} - \frac{1}{p}(R_{x+k+n+1} - R_{x+k+n+p+1}).\right.$$

We now come to the enumeration of the different cases of the payment side of the equation. This we shall divide into two tables, one expressing the terms dependent on the premiums to be paid, the other the returns (where there are any) to be made in the event of no benefits becoming due.

Payment Terms.

1. *Single Premium.* The whole present value of the benefit, σ, paid at once,

$$\sigma D_x.$$

2. *Life Premium.* £ϖ now, and the same at the end of every year during life,

$$\varpi N_{x-1}.$$

3. *Temporary premium.* £ϖ now, and $l-1$ more times, l times in all,

$$\varpi(N_{x-1} - N_{x+l-1}).$$

4_1. *Life Premium, increasing or decreasing by a proportion.* £ϖ now, and $(1 \pm \mu)\varpi$, $(1 \pm 2\mu)\varpi$, &c. in succeeding years,

$$\varpi(N_{x-1} \pm \mu S_x).$$

4_2. When $\mu = 1$, for the *increasing premium*,

$$\varpi S_{x-1}.$$

5. *Temporary Premium, increasing or decreasing by a proportion.* To last only m years, last premium $(1 \pm \overline{m-1}\,\mu)\varpi$,

$$\varpi\{N_{x-1} - N_{x+m-1} \pm \mu\,(S_x - S_{x+m} - mN_{x+m-1})\}.$$

6. *Premium temporary by extinction.* Here $m\mu = 1$, and the extinction takes place after m premiums,

$$\varpi\left\{N_{x-1} - \frac{1}{m}\,(S_x - S_{x+m})\right\}.$$

7. *Arrested proportionally increasing or decreasing Premium.* The premiums of m years are $\varpi, \ldots . \varpi(1 \pm \overline{m-1}\,\mu)$, at which they afterwards remain,

$$\varpi\{N_{x-1} \pm \mu\,(S_x - S_{x+m-1})\}.$$

8₁. *Temporary Premium, continued by proportional increase or decrease.* Here $l+1$ premiums ϖ are to be paid; afterwards $m-1$ premiums $\varpi(1 \pm \mu)$, $\varpi(1 \pm 2\mu) \ldots . \varpi\{1 \pm (m-1)\mu\}$,

$$\varpi\{N_{x-1} - N_{x+l+m-1} \pm \mu\,(S_{x+l} - S_{x+l+m} - mN_{x+l+m-1})\}.$$

8₂. The same *continued to extinction* $(\mu m = 1)$,

$$\varpi\left\{N_{x-1} - \frac{1}{m}\,(S_{x+l} - S_{x+l+m})\right\}.$$

9. *Life Premium, increasing or diminishing by an absolute sum.* Premium ϖ, $\varpi \pm \beta$, $\varpi \pm 2\beta$, &c.

$$\varpi N_{x-1} \pm \beta S_x .$$

10.* *Temporary Premium, increasing or diminishing absolutely.* To last m years, last premium $\varpi \pm (m-1)\beta$,

$$\varpi\,(N_{x-1} - N_{x+m-1}) \pm \beta\,(S_x - S_{x+m} - mN_{x+m-1}).$$

11. *Arrested absolutely increasing or diminishing Premium.* The first m premiums are ϖ, $\varpi \pm \beta, \ldots . \varpi \pm (m-1)\beta$, at which they afterwards remain,

$$\varpi N_{x-1} \pm \beta\,(S_x - S_{x+m-1}).$$

12. *Temporary Premiums continued by absolute increase or decrease.* Here $l+1$ premiums ϖ are to be paid; afterwards $m-1$ premiums $\varpi \pm \beta$, $\varpi \pm 2\beta \ldots . \varpi \pm (m-1)\beta$.

$$\varpi\,(N_{x-1} - N_{x+l+m-1}) \pm \beta\,(S_{x+l} - S_{x+l+m} - mN_{x+l+m-1}.)$$

The following is the table of modes of returning a portion of the premiums.

N.B. These tables do not suffice to calculate the effect of the return of a given proportion of varying premiums. The quantities following are *positive* when put on the benefit side, and negative when on the payment side. It must, of course, be obvious that this is only another table of the values of assurances, described so as to meet the form in which problems are usually given.

Return Terms.

1. *Fixed Return at death.* A fixed sum ϱ returned whenever the death may take place,

$$\rho M_x .$$

2. *A fixed Return at death, if before w years have elapsed,*

$$\rho\,(M_x - M_{x+w}).$$

3. *Return at death of a proportion of fixed premiums.* That is, $\nu\varpi$, if the death take place in the first year, $2\nu\varpi$ if in the second, &c.

$$\nu\varpi R_x.$$

4. *Return at death of a temporary proportion of fixed premiums,* if the death take place within w years.

$$\nu\varpi\,(R_x - R_{x+w} - wM_{x+w}).$$

5. *An increasing or decreasing sum returned at death;* ρ if in the first year, $\rho \pm \theta$, if in the second, &c.

$$\rho M_x \pm \theta R_{x+1}.$$

6. *The same if the death take place in w years,*

$$\rho\,(M_x - M_{x+w}) \pm \theta\,(R_{x+1} - R_{x+w+1} - wM_{x+w}).$$

7. *An arrested increasing or decreasing sum returned at death.* ρ if in the first year, $\rho \pm (w-1)\,\theta$ if in the wth, or any following year,

$$\rho M_x \pm \theta\,(R_{x+1} - R_{x+w}).$$

8. *A fixed sum, ρ, or a fixed proportion of premiums* returned if the life continue w years,* ρD_{x+w}.

A few general rules will be readily collected from the preceding, and may be simply demonstrated. They might be made the foundation of a synthetical view of the subject.

1. Every thing depends on the fundamental calculation of the various cases of the benefit side of the equation.

2. The benefit side of the equation being found for the whole life, that for the same benefit deferred k years is found by writing $x+k$ for x.

3. And that for the same benefit to last n years is found by changing

$$N_x \text{ into } N_{x,\,n},\ M_x \text{ into } M_{x,\,n}.$$

With regard to S_x and M_x the change must be regulated by the following consideration. When their exponent is *one more* than the present age, or the age at a term mentioned in the problem, change S_x into $S_{x,\,n}$, and R_x into $R_{x,\,n}$. But whenever S_x or R_x has the exponent of the present age, or of that at the beginning of a term, change S_x into $S_{x,\,n+1}$, and R_x into $R_{x,\,n+1}$, (compare (page 16,) the transition from 6_1 to 8_1 with that from 6_2 to 8_2). If no simplifications were allowed, that is, if N or M were always retained for the permanent portion of an annuity or insurance, and S or R for the term depending on the value of the incremental portion, the first rule would be sufficient.

4. All the cases are then derived from the following:

N_x on the benefit side of the equation, an annuity of £1, £1, £1, &c.
M_x . . . . assurance of £1, £1, £1, &c.
S_x . . . . annuity of £1, £2, £3, &c.
R_x . . . . assurance of £1, £2, £3, &c.

* For a given year a proportion of the premiums paid by that time is simply a fixed sum.

5. Every formula in which any particular relations exist should be carefully looked at with a view to simplification.

If we wish to make a benefit begin k years *earlier*, we write $x-k$ for x. In the case of an immediate annuity, this is intelligible enough when $k=1$, but N_{x-k} when k is greater than 1, is the impossible quantity of this branch of algebra. Its meaning is as follows: suppose a person has been k years in the enjoyment of a benefit, or of the chance of a benefit, for which he ought to have paid when such enjoyment began; suppose also that, had he died during the k years, the claimant of the payment would have had no means of recovering his rights. According to the principles which regulate these transactions, the holder of the unbought benefit ought to pay the claimant not only all arrears with compound interest, if any, but also compensation for the chance of loss which he has run. The value of such compensation is found by writing $x-k$ for x in the value of the benefit as reckoned from the present time. Thus, $N_{30} \div D_{50}$ is what a person now aged 50 should pay for the past and the future, who has been in unbought possession of an annuity of £1 for 20 years; and $C_{30} \div D_{50}$ is what a person aged 50 should now pay for the unbought chance of having formerly received £1, if he had died between 30 and 31.

This last consideration will be particularly important in its application to the commutation tables for interests certain, as we may thus find the value of all arrears, or may solve a case in which partly arrears and partly prospects are to be valued. For instance, a person engaged to pay a decreasing rent for certain tenements, £a at the end of the first year, $a-h$ of the second, &c., and $a-(n-1)h$ at the end of the nth and last: k years elapse during which he pays no rent, and then his affairs pass into the hands of assignees, who are desirous of paying the arrears and buying the remaining term for one sum. Here at the commencement, the payment side of the equation is σD_x, and the benefit side is $aN_{x,n} - hS_{x+1,n}$: put the last back k years (assuming the age, which is indifferent,* to be k) and we have

$$\sigma = \frac{a(N_0 - N_n) - h(S_1 - S_{1+n} - nN_n)}{D_k},$$

which is the sum to be demanded of the assignees.

It now only remains to show an example of the mode of proceeding with the registered cases.

An assurance, to commence in k years, and to be £A, A+H ....A+$(n-1)$H in the following n years, at which last sum it is arrested, together with an annuity of £a, to begin at the same term, and to last n years, is to be bought by present payment of a sum σ, and also of a premium which is extinguished after k years, or in $k+1$ payments, on condition that the sum σ shall be returned, with simple interest, if the life drop during the k years. Required the first premium ϖ

* In the tables for interests certain it will do equally well to put the payment sid forward k years.

Benefit terms.	The assurance (13), $AM_{x+k}+H(R_{x+k+1}-R_{x+k+n})$	$=V$
	The annuity (5), $a(N_{x+k}-N_{x+k+n})$	$=W$
Payment terms.	The fixed sum σ (1), σD_x	$=X$
	Premium (6), $\varpi\left\{N_{x-1}-\frac{1}{k+1}(S_x - S_{x+k+1})\right\}$	$=Y$
Return terms.	Return of $\sigma(1+r)$ in the first year, &c.	
	$\sigma(1+r)(M_x-M_{x+k})+r\sigma(R_{x+1}-R_{x+k+1}-kM_{x+k})$	$=Z$
	or $\sigma(M_x-M_{x+k})+r\sigma(R_x-R_{x+k}-kM_{x+k})$.	

As no further simplification suggests itself, each term had better be calculated for the particular case wanted: we have then

$$X+Y\varpi-Z=V+W, \quad \varpi=\frac{V+W+Z-X}{Y}.$$

When it becomes necessary to return proportions of decreasing premiums, new tables must be constructed from S_x and R_x, say Z_x and Y_x, so that

$$Z_x = S_x + S_{x+1} + S_{x+2} + \ldots.$$
$$Y_x = R_x + R_{x+1} + R_{x+2} + \ldots.$$

These, divided by D_x, will give the values of the annuity or assurance £1, 3, 6,...., or $n\frac{n+1}{2}$ in the nth year: and the effect of these tables, combined with the others, will be to give the value of an annuity or assurance which is $a+hn+qn^2$ in the nth year.

In Mr. Barrett's original method, which is still followed by some actuaries, are three columns only, answering to D, N, and S, which by aid of the first three formulæ VIII., give C, M, and R. The *great principle* of the method, namely, the formation of tables by which deferred, temporary, and increasing benefits are as easily calculated as those for the whole life, belongs to Mr. Barrett as much as the invention and construction of logarithms to Napier. On the other hand, Mr. Griffith Davies, by the alteration presently noted, and the separate exhibition of M and R (he has not given C, which is of little use in practice, though essential to the theory) has increased the utility and extended the power of the method to an extent, of which its inventor had not the least idea; and has all the rest of the claim in the matter which is made for Briggs in the adaptation of logarithms to practical use. Nor must it be forgotten, that in all probability this most expeditious mode of conducting operations would not have been now in existence, but for the sagacity of Mr. Baily, who, as we have seen, saw further into its merits than the Royal Society. In Mr. Barrett's form, there are three columns, A, B, and C: and A_x is not $a_x v^x$, but $a_x(1+r)^{w-x}$, where w is the greatest age any individual can attain. Also B_x is not $A_{x+1}+\ldots$, but $A_x+\ldots.$, so that the value of an annuity on a life aged x is $B_{x+1} : A_x$. Again, C_x is $B_x + B_{x+1} + \ldots.$ The following comparisons may be useful to those who are habituated to Barrett's original form, remembering that C_x now means Barrett's third column, and not what it has hitherto stood for :

$$A_x = D_x v^{-w}, \qquad D_x = A_x v^{w},$$
$$B_x = N_{x-1} v^{-w}, \qquad N_x = B_{x+1} v^{w},$$
$$C_x = S_{x-1} v^{-w}, \qquad S_x = C_{x+1} v^{w}.$$

Since, then, Barrett's form is that of Mr. G. Davies multiplied by a constant factor, the former are also *commutation* tables, using the life a year older than the given life in the second and third columns.

In conclusion, we may mention that when a whole table is to be calculated, it may happen that it is better to dispense with the assurance columns by means of VIII. Thus in the case of the premium of assurance for a term of years (as noted by Mr. G. Davies)

$$\frac{M_x - M_{x+n}}{N_{x-1} - N_{x+n-1}} \text{ is not so convenient as } v - \frac{N_x - N_{x+n}}{N_{x-1} - N_{x+n-1}}.$$

The only works of which the writer is aware, in which the preceding method, whether called by the name of Barrett or Davies, is treated, are the Appendix to Mr. Baily's Treatise on Life Annuities, &c.; the French translation of the same; Mr. G. Davies' work, already cited; a Note in the Appendix to Mr. Babbage's Treatise on Life Assurance; and the treatise on Life Annuities, &c. by Mr. Jones, now in course of publication in the Library of Useful Knowledge.

A. DE MORGAN.

University College, London,
October 1, 1839.

Since writing the above, it has struck me that it would be more convenient to make the calculations in page 6 by writing the logarithms of a_x and $a_x - a_{x+1}$, one above, and the other below, the logarithm of v^x, than by writing both of the former below the latter.

II.—HISTORICAL NOTES ON ALMANACS.

Last year we laid before our readers notices of the most celebrated almanacs that were published prior to the Reformation. We now take a wider field, and intend to skim over the progress made in the same ground during the two succeeding centuries.

The almanac of our ancestors was completely *sui generis;* it was not sufficient for one to know the seasons, weathers, and predilections of man and beast, but dive he must into the coming political events, and discover what—

> "Hairy meteors may pronounce
> The fall of sceptres and of crowns."

It was necessary that the whole chain of events should be laid before him as far as the combined systems of prophecy could accomplish the work. Since, therefore, modern prophecy is so closely connected with the *et cætera* of almanacs, it may not be irrelevant if we venture to make a few observations on this curious and interesting branch of inquiry, before we enter into the heart of our intended disquisition.

The pretended source of fore-knowledge was generally through the agency of conjured spirits. The following is translated from

Printed in Dunstable, United Kingdom

91220089R00020